Just Bugs

Learning the Short U Sound

Jeff Jones

Phonics
for the
REAL World™

Rosen Classroom Books and Materials™
New York

We just like bugs.

We put our bugs in a jug.

One bug is black.

One bug is red.

One bug is green.

One bug is brown.

One bug has wings.

One bug can buzz.

One bug lights up.

19

Bugs are fun!

Word List

bug
buzz
fun
jug
just
up

Instructional Guide

Note to Instructors:
One of the essential skills that enable a young child to read is the ability to associate letter-sound symbols and blend these sounds to form words. Phonics instruction can teach children a system that will help them decode unfamiliar words and, in turn, enhance their word-recognition skills. We offer a phonics-based series of books that are easy to read and understand. Each book pairs words and pictures that reinforce specific phonetic sounds in a logical sequence. Topics are based on curriculum goals appropriate for early readers in the areas of science, social studies, and health.

Letter/Sound: short u – Say the following words and have the child define them and/or use them in oral sentences: *duck, tug, bus, bumpy, cup, drum, hug, fun, gum, jump*. List the words on a chalkboard or dry-erase board. Ask how all the words are alike. Have the child underline the **short u** in each word.
- Have the child find the longest word in the list, the words that rhyme, the word that names something that flies, etc.

Phonics Activities: On the chalkboard or dry-erase board, write one-syllable words with **short u**, such as: *up, cup, pup, us, bus, fuss, Gus, muss, bug, mug, tug*. Encourage the child to use consonant sounds as well as **short u** sounds as they decode the words. Have them underline the **short u** in each word.
- Provide the child with a response card bearing the word *buzz*. Pronounce several words with short vowel sounds, such as: *back, hop, dish, luck, stick, doll, slip, sun, cup,* etc. Have the child hold up their card and make a buzzing sound when they hear **short u** words. As they respond, write words on the chalkboard or dry-erase board and have the child underline the **short u**.
- Say the following words and write them on the chalkboard or dry-erase board: *sun, dust, pup, bun, tub, jug, much*. Have the child tell one way that all the words are alike. Ask them to think of words that rhyme with each and have the **short u** sound *(fun, rust, cup, run, cub, rug, such)*. Have the child underline the **short u** in each word.

Additional Resources:
- Dussling, Jennifer. *Bugs! Bugs! Bugs!* New York: DK Publishing, Inc., 1998.
- Kalman, Bobbie. *Bugs & Other Insects*. Madison, WI: Demco Media, Limited, 1994.
- Reid, Mary, and Betsey Chessen. *Bugs, Bugs, Bugs!* New York: Scholastic, Inc., 1997.
- Scarborough, Sheryl. *About Bugs*. San Francisco, CA: Treasure Bay, Inc., 1999.

Published in 2002 by The Rosen Publishing Group, Inc.
29 East 21st Street, New York, NY 10010

Copyright © 2002 by The Rosen Publishing Group, Inc.

All rights reserved. No part of this book may be reproduced in any form without permission in writing from the publisher, except by a reviewer.

Book Design: Haley Wilson

Photo Credits: Cover © C. C. Lockwood/Animals Animals; p. 2 © David McGlynn/FPG International; p. 3 © Mark Hunt/Index Stock; p. 5 © Larry Barnes/Black Star Publishing/PictureQuest; pp. 7, 13 © Donald Specker/Animals Animals; pp. 9, 11 © Bill Beatty/Animals Animals; p. 15 © Buddy Mays/International Stock; p. 17 © J. R. Williams/Animals Animals; pp. 18, 19 © James E. Lloyd/Animals Animals; p. 21 © Michael Bisceglie/Animals Animals.

Library of Congress Cataloging-in-Publication Data

Jones, Jeff (Jeff W.)
  Just bugs : learning the short U sound / Jeff Jones.
    p. cm. — (Power phonics/phonics for the real world)
  ISBN 0-8239-5911-2 (lib. bdg. : alk. paper)
  ISBN 0-8239-8256-4 (pbk. : alk. paper)
  6-pack ISBN 0-8239-9224-1
  1. Insects—Juvenile literature.   [1. Insects.]   I. Title.
  II. Series.
  QL467.2.J66 2001
  595.7—dc21

                              00-013129

Manufactured in the United States of America